TYPES OF AI

NANCY DICKMANN

BROWN BEAR BOOKS

Published by Brown Bear Books Ltd
4877 N. Circulo Bujia, Tucson, AZ 85718
USA
and
Studio G14, Regent Studios, 1 Thane Villas,
London N7 7PH, UK

ISBN 978-1-83572-045-5 (library binding)
ISBN 978-1-83572-051-6 (paperback)
ISBN 978-1-83572-057-8 (ebook)

Library of Congress Cataloging-in-Publication Data available on request

Text: Nancy Dickmann
Consultant: Matthew Lugg
Design and Illustrations: Square and Circus
Design Manager: Keith Davis
Children's Publisher: Anne O'Daly

Picture Credits
The photographs in this book are used by permission and through the courtesy of:
Cover: Freepik.com: **Interior:** Brown Bear Books: 12–13; Shutterstock: Face Stock 6–7, Gorodenkoff 4-5, Insta_photos 10–11, Rapeepat Pomsipak 8–9, PR Image Factory 14–15, Aleksandr Rybaiko 18–19, Luri Stepanov 16–17, Miljan Zivkovic 20–21.
Artwork: Freepik.com.
All other artwork and photography
© Brown Bear Books.

Brown Bear Books has made every attempt to contact the copyright holder. If you have any information about omissions please contact: licensing@brownbearbooks.co.uk.

Websites
The website addresses in this book were valid at the time of going to press. However, it is possible that contents or addresses may change following publication of this book. No responsibility for any such changes can be accepted by the author or the publisher. Readers should be supervised when they access the Internet.

Words in **bold** appear in the Glossary on page 23.

Manufactured in the United States of America
CPSIA compliance information: Batch#AG/5666

CONTENTS

COMPUTERS THAT LEARN

We use computers every day. How do you use them? You might do homework or watch videos. You can search things up on the internet. You can make music or edit photos. There are computers at schools. Many homes have them too. They are good at tasks like these.

ARTIFICIAL INTELLIGENCE

Some computer **programs** go a bit further. They are designed to think and learn like humans do. This is artificial intelligence. We call it AI for short. There are many different kinds of AI. They do different tasks. They are built and trained in different ways.

Your brain is good at learning. Some computers can learn, too.

Your Amazing Brain

The human brain can do a lot. Scientists are working on AI that can do the same things.

Take in information about the world

Understand and use language

Look for **patterns**

Store memories and facts

Recognize things you haven't seen before

Solve problems and puzzles

TALKING TO AN AI

Too Hard?

Sometimes AI doesn't understand what you say. It knows your language. But it might have trouble with your accent. It needs more training on different accents.

When you talk to someone, you speak the same language. The same is true for AI. Many computers only understand special computer languages. But you can "talk" to some AI. You type using your own language. The AI understands. It replies in the same language.

SPEECH AND TEXT

Some AIs understand **text**. They might be trained to read several different languages. Other AI goes a step further. It can understand spoken words. You give it a voice command. Then it turns this into text. Many people write text messages this way.

Some homes have smart speakers. You can ask them to play music or turn on the heating.

Learning to Speak

An AI has to be taught to understand language. Here is how it works.

The AI is given lots of text.

Each piece of text has a label. It tells the AI what it is.

The AI looks for patterns in the text.

It learns what words mean.

It learns how sentences are put together.

Now it can write its own sentences.

GENERATIVE AI

Some AI is designed to create content. We call it generative AI. The content could be text or images. It could be music or animation. To do this, AI studies lots of different examples. It figures out what they have in common. Then it makes its own version.

TRULY CREATIVE?

Artists and authors use their creativity. They can create fantasy worlds or new styles. They might be inspired by other people's work. Then they put their own spin on it. AI is not as good at being creative. It relies more heavily on what it's seen. The content it produces might sound good. But it isn't very original.

Prompts

You need to tell an AI what you want. This is called a **prompt**.

- Do you want text, music, or a picture? Most AIs can only do one kind of thing.
- What's the subject of the work?
- What style do you want?
- How long should it be?

Prompts should be clear and specific. They help you get what you want from the AI.

It might take you an hour to write a good story. An AI can write one in seconds.

WORKING WITH WORDS

Many people use **chatbots**. These are a kind of generative AI. You type a question or prompt. The AI gives you an answer. These AIs use their language training. They understand what you are asking. They give the best reply they can.

IMPROVING TEXT

These chatbots can write new text. They can also make your own writing better. Chatbots check grammar and spelling. They can rewrite text to make it sound better. They use different words. They break up sentences. These changes make the text clearer.

You can talk to a chatbot on a phone or a laptop. You connect using the internet.

Making Things Up

AI doesn't always give the right answer. It is just making its best guess. It might make the answer up, even if it knows the right one. You should always check it somewhere else.

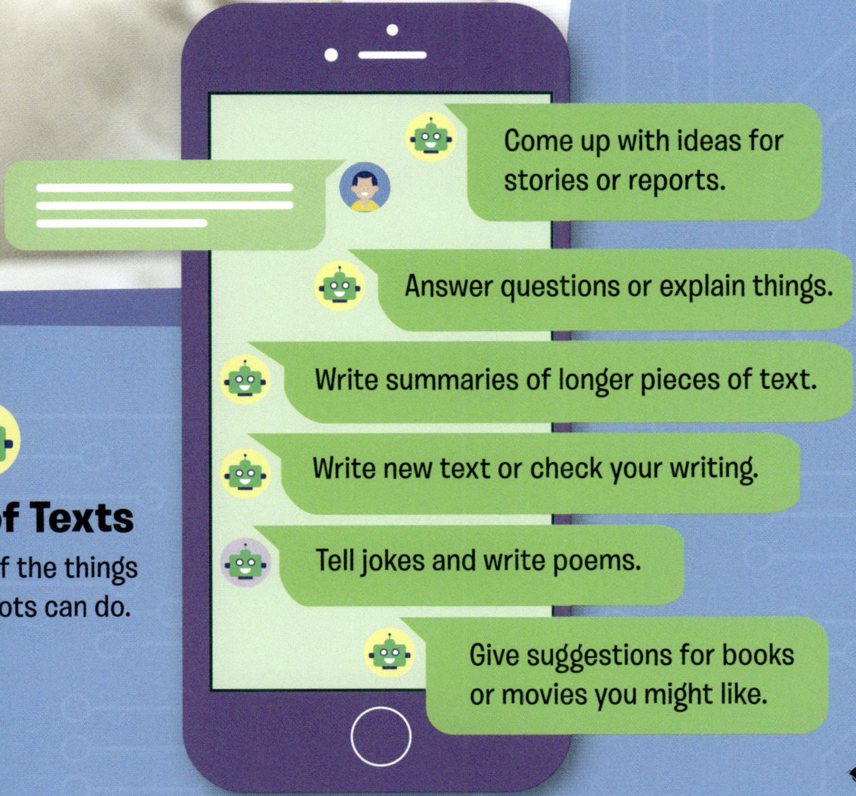

All Kinds of Texts

Here are some of the things that text chatbots can do.

Come up with ideas for stories or reports.

Answer questions or explain things.

Write summaries of longer pieces of text.

Write new text or check your writing.

Tell jokes and write poems.

Give suggestions for books or movies you might like.

CREATING ART

Artists make beautiful paintings. It might take days, or even weeks. They use their skills. They also use their imagination. Some AI can create images too. The images might look like photos. They can also look like paintings. An AI can make one in seconds.

MUSIC AND VIDEO

Some AI makes music. You tell it what style you want. You choose the instruments. You can ask it to sound like your favorite artist. Other AI creates animations. Some turn videos into animations. Others start with a script.

An AI made this image in seconds. It looks mostly real. But some small details are not quite right.

How It Works

Als produce different kinds of art. But they are trained in similar ways.

You feed the AI lots of examples. These could be paintings or song recordings.

Each example has a label. It is called training **data**.

The AI learns from this data. It looks for patterns, such as rhythms.

A user gives the AI a prompt.

The AI searches its **memory**. It looks for similar examples.

The AI uses the examples. It makes its own version.

COMPUTER VISION

You see things with your eyes. Your brain tells you what they are. Say that you see a car. You've seen lots of cars before. This one might look a little different. Your brain compares it to memories of other cars. It recognizes it. It knows it's a car.

COMPUTERS THAT SEE

Human brains are really good at this! It's trickier for computers. But AI is helping them get better. We train programs to recognize images. This is called computer vision. It helps them identify faces or things in pictures.

See, Then Act

Recognizing a picture is the first step. An AI can act on what it sees. Some phones use facial recognition. An AI recognizes the owner's face. Then it opens the phone.

Some cars drive themselves. They must be able to spot people on the road.

Learning to See

We train an AI on images. Each has a label. The AI learns what they show. It applies this to what it sees.

Some doorbells have a camera. It senses motion.

The camera records something moving.

AI helps it recognize what it sees.

Is it a cat or a car? Is it a person? Could it be a burglar?

The AI decides if there is a threat. Then it can send an alert.

FINDING PATTERNS

There is a lot of data in the world. Data can be facts and figures. It can be images, such as pictures of space. We need to make sense of all this data. Computers and AI can help us. They can find patterns. And they work much faster than we can!

MACHINE LEARNING

We give an AI data that it can learn from. It needs a lot! The AI's program shows it how to **analyze** the data. It learns to see patterns. Then it can sort and organize data. It can see trends. It can even make predictions! This is called **machine learning**.

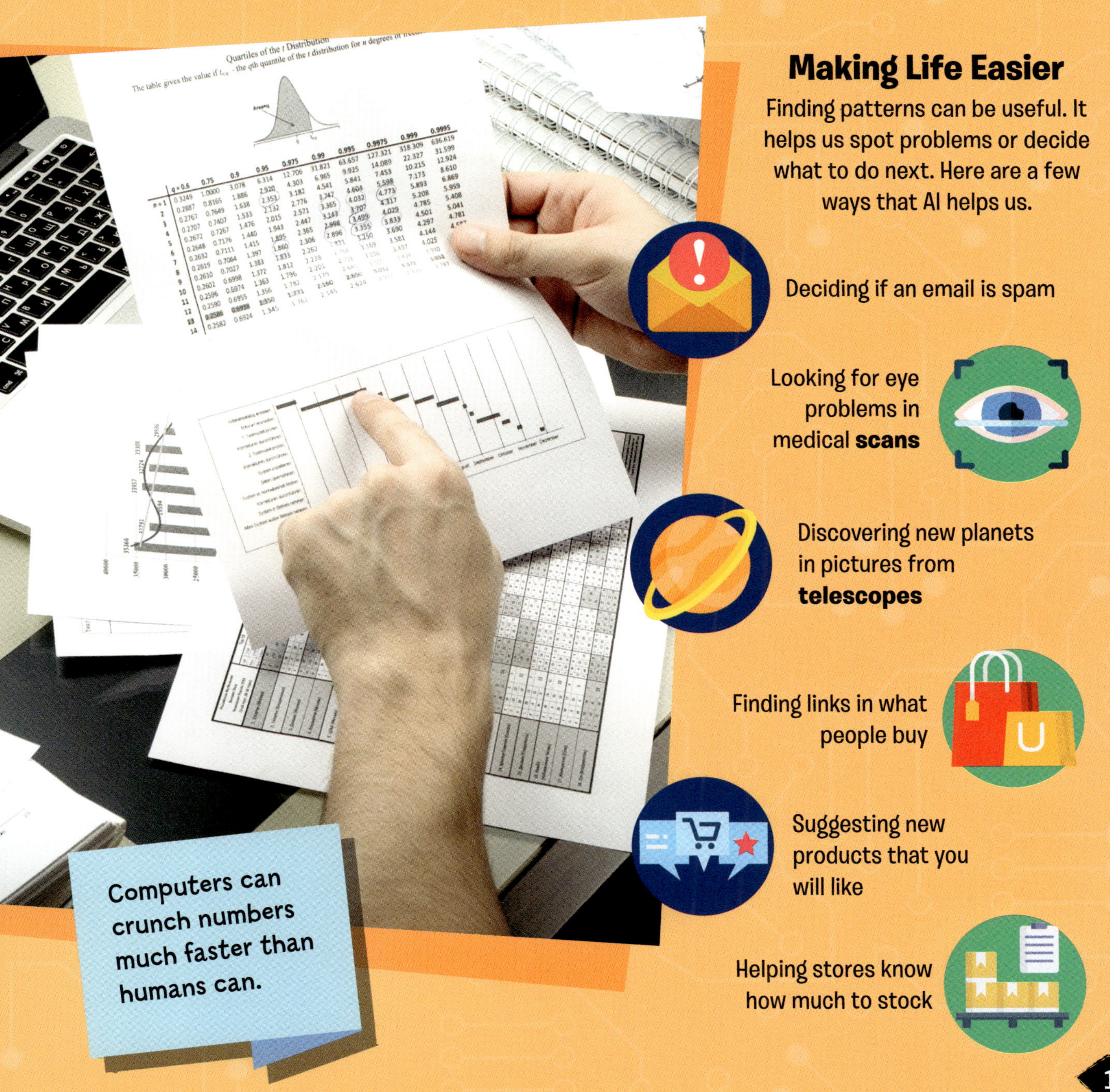

Making Life Easier

Finding patterns can be useful. It helps us spot problems or decide what to do next. Here are a few ways that AI helps us.

- Deciding if an email is spam
- Looking for eye problems in medical **scans**
- Discovering new planets in pictures from **telescopes**
- Finding links in what people buy
- Suggesting new products that you will like
- Helping stores know how much to stock

Computers can crunch numbers much faster than humans can.

MAKING MACHINES WORK BETTER

Each AI is different. They are all designed for a specific job. Some AI is meant to help machines work better. They take in data from the machines. Then they make decisions. Could the machine work more efficiently? Does it need fixing or upgrading?

HOW IT WORKS

Machines such as airplanes rely on engines. These engines often have sensors. They measure heat and vibrations. An AI can look at this data. It **predicts** when problems are coming. Then the engine can be fixed before it breaks.

AI helps farmers predict the best time to harvest their crops.

Keeping the Lights On

We track how much electricity is being used. AI can help analyze this data. It predicts times when more electricity will be needed. We can make sure there is enough.

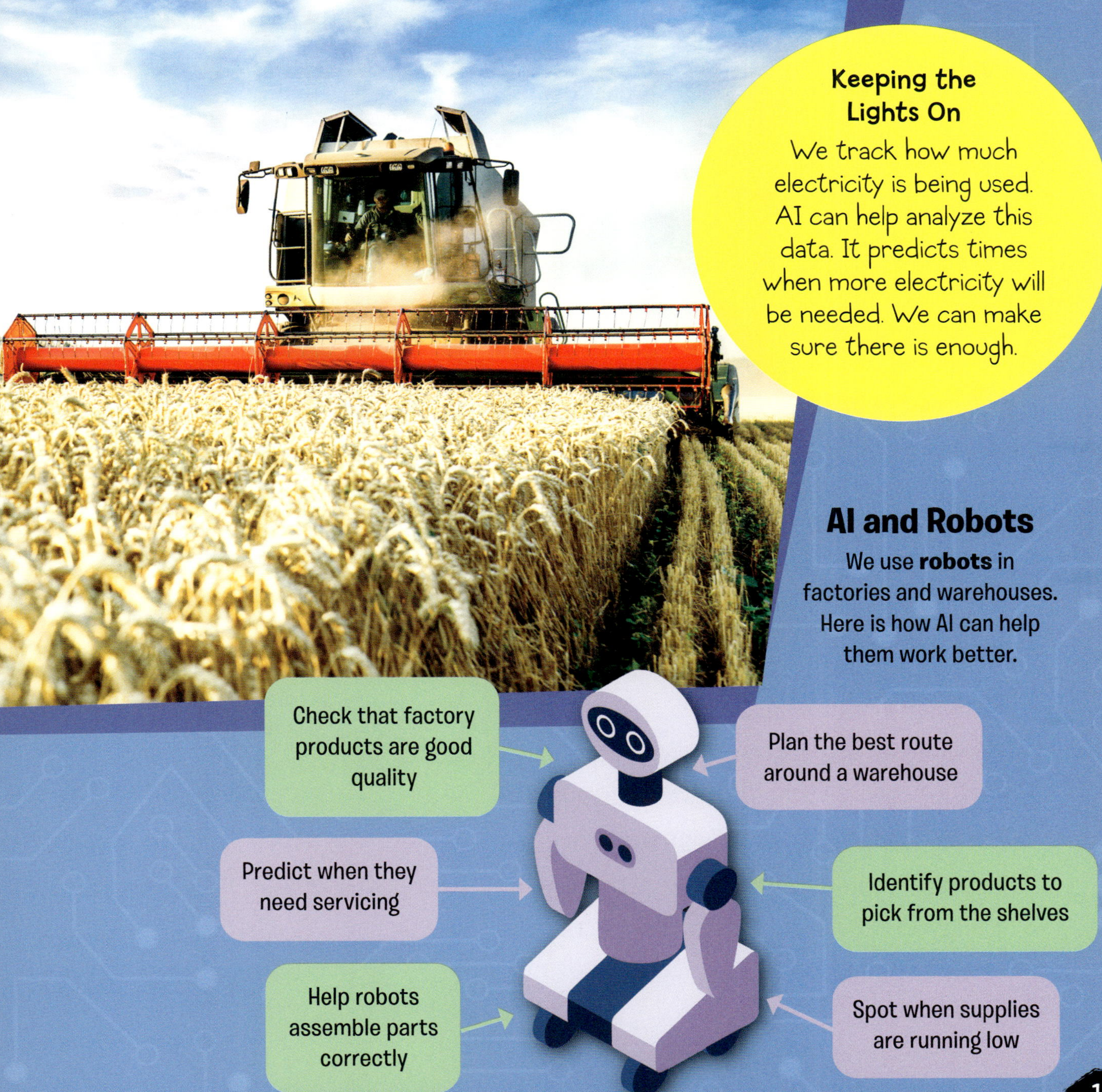

AI and Robots

We use **robots** in factories and warehouses. Here is how AI can help them work better.

PLAYING GAMES

Many AIs do serious jobs. They help doctors. They keep factories running. But AI can also play games! Some AI is trained to play the games we like. AI can often beat human players. It crunches the numbers to choose the best move.

In a game like chess, you have to think several moves ahead. AI is good at doing this.

ALL KINDS OF GAMES

There are AIs that play board games, such as checkers or chess. Others play card games. Some play video games like Minecraft. AI can also solve puzzles like crosswords or sudoku. Each AI is trained for a particular game.

Who Wins?

When it comes to playing games, AI has many advantages. Here are a few.

QUIZ

How much have you learned about the different types of AI? It's time to test your knowledge!

1. What kind of AI produces text, pictures, or music?

a. artistic AI

b. generative AI

c. imagination AI

2. What do we add to data that is given to an AI to train it?

a. labels

b. spreadsheets

c. cookies

3. What is it called when AI can recognize objects or photos?

a. image finding

b. automatic looking

c. computer vision

4. Which of these games is AI good at playing?

a. chess

b. basketball

c. hopscotch

The answers are on page 24.

GLOSSARY

analyze to examine something in detail in order to find patterns or explain it

chatbot a computer program designed to have conversations with humans

data information that is stored or used in a computer, in the form of a series of ones and zeroes

emotion a feeling such as joy, anger, or fear

machine learning a method that lets a computer learn to do a task by analyzing lots of data

memory the ability to store and recall events or information from the past

pattern an arrangement of similarities or trends between different items in a set

predict to make an educated guess about what will happen

program a set of coded instructions for a computer to follow

prompt an instruction given to a computer or AI to tell it what to do

robot a machine with moving parts that is programmed to do a job

scan an image taken of part of the body to help doctors spot injury or disease

telescope a scientific tool for looking at things that are far away, such as in space

text data in the form of written words

FIND OUT MORE

Books

AI Basics. Elsie Olson, Lerner Publications, 2025.

Artificial Intelligence. Julie Murray, Abdo Books, 2021.

How AI Works. Lisa Idzikowski, Lerner Publications, 2025.

Websites

www.bbc.co.uk/newsround/49274918

kids.britannica.com/students/article/ChatGPT/636542

www.softwareacademy.co.uk/ai-for-kids/

INDEX

Answers: 1. b; 2. a; 3. c; 4. a